Off-Grid Solar Power

A Step-by-Step Guide to Make Your Solar Power System for RVs, Vans, Cabins, Boats and Tiny Homes.

Alexander Ketch

© **Copyright 2020 - All rights reserved.**

The content contained within this book may not be reproduced, duplicated or transmitted without direct written permission from the author or the publisher.

Under no circumstances will any blame or legal responsibility be held against the publisher, or author, for any damages, reparation, or monetary loss due to the information contained within this book, either directly or indirectly.

Legal Notice:

This book is copyright protected. It is only for personal use. You cannot amend, distribute, sell, use, quote or paraphrase any part, or the content within this book, without the consent of the author or publisher.

Disclaimer Notice:

Please note the information contained within this document is for educational and entertainment purposes only. All effort has been executed to present accurate, up to date, reliable, complete information. No warranties of any kind are declared or implied. Readers acknowledge that the author is not engaged in the rendering of legal, financial, medical or professional advice. The content within this book has been derived from various sources. Please consult a licensed professional before attempting any techniques outlined in this book.

By reading this document, the reader agrees that under no circumstances is the author responsible for any losses, direct or indirect, that are incurred as a result of the use of the information contained within this document, including, but not limited to, errors, omissions, or inaccuracies.

ISBN: 9798674988236

Table of Contents

Table of Contents..3
Introduction..5
Chapter 1: Going Off-Grid..8
    Why Solar Power?...9
    How Does an Off-Grid Solar Power System Work?................12
    The Lifestyle Benefits of Using Off-Grid Solar Power...............15
Chapter 2: What to Consider Before You Start.................................18
    Calculating the Amount of Energy Your Home Will Require..................20
    Materials and Tools for Building an Off-Grid Solar Power System.........29
    Calculation of the wire section (12V DC).....................................32
    Calculation of the wire section (120V/220V CA).......................35
    Choice of Battery..36
    Calculating Costs..42
Chapter 3: Building Your Off-Grid Solar Power System.................43
    DIY: Solar Power for Tiny Homes and Cabins............................43
        Key Questions to Ask Before Building....................................44
        Calculations for Your Off-Grid Tiny Home or Cabin............45
    DIY: Solar Power for RV's and Boats..47
        Key Considerations Before Building in Your RV or Boat...................48
        Calculations for an Off-Grid RV or Boat................................50
Chapter 4: Tips and Common Mistakes When Building an Off-Grid Solar System..52
    Avoid These Ten Costly Mistakes..54

Maintaining Your Solar System in Every Season..................................58

Conclusion...61

Your Gateway to Freedom...63

References..66

Introduction

A few decades ago, solar power was hardly spoken of—at least in the public domain. I remember the day when I first heard of a solar panel. The concept of a board that you could place on your rooftop, and convert energy from the sun into electricity was remarkable! I never understood how no one had thought about inventing this device sooner. The more I learned about solar power systems, the more obvious the solution of solar-powered electricity became. The less natural I found the use of fossil fuels in generating electricity.

This is because fossil fuels are a non-renewable energy resource, which means that when all of the reserves have been depleted, we will be left in the dark. Another pet peeve regarding fossil fuels—which as a society we can no longer ignore—is how these fuels release gases into the air and in our water systems, which contaminate our environment, causing cancerous diseases that become apparent much later in life. It is no longer viable for us to continue using harmful energy to power our homes and expect to live a healthy lifestyle. The revelation of solar power is our ticket to regaining control over what we allow into our lives. It comes from a free natural resource that is reliable and generous in allowing us to have as much electricity as our homes require.

Solar power systems are investments that we make once and reap a hundredfold return from for the rest of our lives. They are particularly beneficial for people who have decided to live off the grid. Living "off-grid" is a lifestyle decision that many people make, which involves living in locations or homes that are not maintained by the state's primary electric utility grid. These people live in self-sufficient households surrounded by plenty of nature and full access to natural resources like water and the sun, which help sustain their livelihoods.

Solar power systems are useful in feeding electricity within remote homes and allowing the household to save a lot of money. Relying on the sun for electricity is far more affordable than installing an electric utility pole near the home or running cables from the closest main utility grid access point. In other words, solar power provides an opportunity for separate households to become completely self-sufficient, generating their own heat to cook food, take warm showers and fulfill any other useful electrical needs that the household would enjoy.

In this book, I intend to give you the most relevant knowledge and practical guidance in constructing your solar power system for remote homes such as tiny houses, cabins, RV's, and houseboats. Creating your own off-grid solar power system may seem like an overwhelming task requiring a science degree; however, this is not true. All that you will need is some basic math and an open mind. I will provide you with some simple designs on how to create a solar power system depending on the type of house you live in, but your design will end up customized to your home and specific to daily electricity requirements.

Many people may be weary of going completely off-grid out of fear of the costs involved in owning one of these systems. While off-grid solar power systems are a lot more expensive than on-grid solar power systems (which still rely on the main grid) the investment is still worthwhile. One brilliant way to make your system a lot cheaper, is to reduce your power demands by switching to energy-saving electrical appliances, machines and accessories. Using greener and more energy efficient products will reduce the load that your solar power system has to carry, which allows you to build a much smaller system that requires fewer parts.

In this book, I will guide you in making smart purchasing decisions when buying components of your off-grid solar power system. I will also educate you on the different kinds of products that are on the market which can meet all of the power requirements, depending on the size of your system. Another important element of the book will be in showing you how to perform the necessary calculations which will ultimately determine (with the highest amount of accuracy) how many parts you need and at what voltage. These calculations will become your blueprint in strategically designing your system on paper before purchasing any solar products. Lastly, I will share with you some common mistakes that many people make when installing their solar power system and ways of avoiding these blunders. Remember that mistakes will cost you more money and therefore it is advised that you take your time in understanding the system—and how to take care of it—before undergoing your project.

After reading this book, you will feel empowered to take on a unique DIY project that will sustain you and your family for many years to come. You would have learned about the different components of a solar system and how to assemble these components at home (perhaps it may also inspire you to find other smaller projects that would benefit from the free supply of solar energy). Nonetheless, the biggest takeaway you will receive from reading this book will be knowing just how rewarding it is to build a custom source of power and save money in the process!

Chapter 1: Going Off-Grid

The world that we live in today does not resemble the one that we grew up in. My childhood was full of rich experiences involving playing outside in the dirt, eating food that was grown from our backyard, and using my hands to create, learn, and serve others. However, society has drastically changed since then. Nowadays, we live in communities controlled by a consumerist agenda where we are encouraged to consume more than we need. As a result of this culture of consumption, many people are deep in debt and cannot find peace from this lifestyle. Furthermore, the promotion of consumption has led to saddening levels of pollution caused by the greed of industries, whose only mission is to make more and give less.

Our environment is crying out for the life of yesteryear, an existence that did not require so much human intervention and nature to provide for our needs. More than ever, we see people who share the same cries and desire to go back to the basics. A life spent living away from the concrete jungle and immersed in the beauty of our natural surroundings is beginning to appeal to many. Some have gone as far as living in remote locations where they are independent of any government support or intervention. We call this kind of lifestyle an "off-grid" or off the grid lifestyle because essentially, the individual cannot be pin-pointed on a map. They are not connected to any governmental power utilities where they can be sent a bill at the end of the month for the service.

Off-grid living empowers us to return to a simple life where we are sustained by the same environment we seek to protect. Most people who decide to live off the grid are naturalists who enjoy conserving nature, protecting animals, reducing waste, recycling, and being frugal with cash flow (and where their money is spent). Many of the off-grid homes across the country are completely self-sufficient because they do not rely on the state's services or products for assisted living. This is possible because many of these households rely heavily on the free natural resources within the surroundings for many of their needs.

For instance, the tenants grow their own food instead of purchasing from a supermarket, and they have access to plenty of freshwater from the nearby streams.

Why Solar Power?

Each day, the sun releases a vast amount of energy, which is referred to as solar energy. In a day, the sun radiates more energy than the world will use in one year. All of this energy comes from within the sun itself; the sun is the largest star in our solar system, and it consists of primarily hydrogen and helium gas. The solar energy that is released from the sun is made through a process known as nuclear fusion. Once the energy has been made, it takes only eight minutes to travel 93 million miles and reach the Earth (the traveling speed is approximately 186,000 miles per second).

Even though we only see a portion of the solar energy released from the sun, the amount that we receive in an hour will be more than enough to power the Earth. We classify solar energy as a renewable energy source for this reason; solar energy is a natural resource that will continue to replenish our planet for millennia to come. Solar energy is useful in our society for heating water or powering buildings with electricity. In the United States, over 2 million solar power installations have been successfully made, and this figure is forecasted to rise to over 3 million installations in 2021, according to Wood Mackenzie (2019).

Heating with solar energy is not as straightforward as you might think. When the energy reaches the Earth, it is spread out across a large area, making it difficult to gather a significant amount of energy in one location at any particular time. Furthermore, some regions in the world do not experience long sun exposure or have fluctuating weather conditions, which further makes it challenging to collect solar energy. To assist individuals in concentrating solar energy inside buildings, households typically use a solar collector. Think of a parked car on a sunny day as an example. As the sun rays permeate through the car window, they are absorbed in the chair covers, dashboard, and car floor. The absorbed sun rays are transformed into heat, and the driver will feel this heat as they enter the car. Therefore solar collectors are plates that absorb sunlight and transform it into heat energy.

Solar Space Heating

Many households will use solar energy to heat the inside of the home. Some homes are constructed to let in as much sunlight as possible through the structure of the building and, in essence, function like a big solar collector. The sunshine makes its way inside the home through large windows and effectively heats the walls and floors. The beauty of these solar homes is that the sunlight transfers light turning into heat inside the home and the heat does not have a pathway to escape. This type of home does not depend on large equipment such as pumps or blowers for heat which allows for affordable maintenance of the home.

Solar Water Heating

Solar energy is also used for heating water. Every household can benefit from having access to hot water to perform some of the household tasks such as cooking, taking showers, washing dishes and clothes. You might be interested to know that washing clothes alone accounts for the second-largest energy cost in a household. This is due to most families using washing machines to clean their clothes, which needs a lot of energy. Installing solar water heaters could potentially reduce your water heating bill by as much as 50 percent. These solar water heaters function similarly to solar space heaters in the sense that it requires a solar collector that heats the water in a water tank or geyser, and this hot water is sent through pipes into taps throughout the home.

Solar Electricity

Solar energy is also a fantastic substitute for generating electricity. This electricity can be used to power an entire home. There are two ways electricity can be made from solar energy through photovoltaics (PV) or solar thermal systems. Photovoltaics is a big word that can be broken down into two parts, *photo* which means light and *volt*, which is the measurement of electricity. Some of the products that you are already familiar with are powered by PV cells (these products include solar-powered toys, calculators, or roadside telephone boxes).

However, PV cells are known to supply energy to anything powered using batteries or else electric power.

Electricity is generated when solar energy from the sun reaches the PV cells, which causes the electrons to move around and create an electric current. Similarly to PV cells, solar thermal systems (also referred to as concentrated solar power) can generate electricity from solar energy—however, it is done in a slightly different way. Solar thermal systems make use of solar collectors with a mirrored surface that is useful in concentrating sunlight onto a receiver that will effectively heat a liquid. The heated liquid is then used to produce steam, which subsequently generates electricity.

How Does an Off-Grid Solar Power System Work?

The idea of running your remote home with electricity not regulated by utility companies is an appealing one. Many homeowners are looking for the independence and sustainability that off-grid solar power systems can provide. This technology is now readily available and affordable for many households to utilize. The journey of how solar energy is translated into electricity is not a complicated one to understand. Below I will provide you with a detailed summary of how off-grid solar systems work so that you can finally realize the significance of those shimmering glass panels commonly seen on rooftops.

Step 1: Sunlight will Activate the Solar Panels

Solar panels are built with a layer of silicon photovoltaic cells, a metal frame, a clear glass casing, which is surrounded by a unique film, as well as wiring. Many households prefer to group more than one panel in a series on rooftops to collect as much energy as possible. The photovoltaic cells are responsible for absorbing the maximum amount of sunlight they can receive during the day.

Step 2: The PV Cells Produce an Electric Current

Within each PV cell, you will find a thin semiconductor wafer that is constructed using two layers of silicon. One layer of this wafer is positively charged, and the other is negatively charged, effectively creating an electric field. When the sun rays strike a PV cell, energy is produced within the cell and allows electrons to come apart from atoms found within the semiconductor wafer. These loosened electrons, now free to move around, are powered by the electric field that surrounds the wafer, and this power is what creates an electric current.

Step 3: The Electric Current is Converted

Even though the solar panels have successfully turned sunlight into an electric current, it still has to convert this current into one that can be used to power your home. The electric current, which is initially produced by solar panels, is known as direct current (or DC) electricity; however, to turn on the switches at home and receive light, you will need to convert it to alternating current (or AC) voltage. This process is made possible by using a technological device known as an inverter. In the modern solar power systems used today, you can choose to use one inverter for an entire system or a few singular microinverters propped behind the solar panels.

Step 4: The AC Electricity Powers your Household

Once the inverter has converted a DC electric current to an AC electric current, the current will flow through the home's electrical panel (the main electricity switchboard). It is thereby distributed throughout the home, giving power to all electrical appliances. This type of electricity generated is not different from the electricity that you receive from the utility company. And you do not have to make any changes to the home setup.

Step 5: Batteries Store Excess Solar Energy

The main difference between on-grid and off-grid solar power systems is how excess solar energy is stored. In on-grid solar power systems, any solar energy that you do not use is fed into the utility company's electricity grid to give credits (reduce monthly billing). On the other hand, off-grid solar power systems store the excess solar energy in battery storage systems. The battery systems you use must be large enough to store electricity to power the home for two days or more.

During the night or on days when there is no sun, the electrical appliances will draw power from the electricity stored within the batteries. However, sometimes the batteries are not fully charged, or they cannot produce enough electricity to power your home through extreme winter conditions. In cases like these, you will need to include a backup generator in the system that is large enough to provide backup power in your home and charge the batteries at the same time.

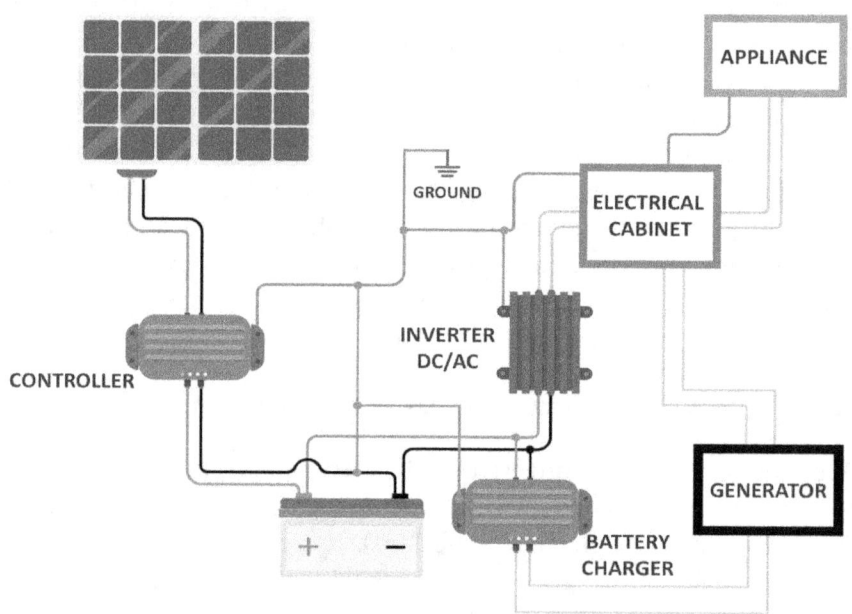

The Lifestyle Benefits of Using Off-Grid Solar Power

Taking a household off the grid using solar power and battery storage is the most effective cost-saving and environmentally conscious decision you can make in 2020 and going forward. I find that even urban dwellers near state electricity supply are opting to become completely energy self-sufficient. I believe that the increase in education surrounding solar power has convinced many households of the immense benefits of using the abundant supply of solar energy for electrical needs. Nonetheless, it does not take too much convincing to see the many advantages of adopting this kind of technology as part of the norm.

1. Avoid power outages.

We have all been in this situation: cooking or listening to the radio, and all of a sudden, without any notice, the power goes out, and you are not entirely sure when it will come back on. Going without lights or working plugs for electrical appliances can be a frustrating experience. The power outage could be a result of extreme weather conditions that damage power lines and other electrical equipment used to feed electricity. Fortunately, off-grid solar systems allow us to store electricity in battery systems so that when blizzards or torrential rains come, you are never without power or heat.

2. Effectively reduce electricity costs.

In many parts of the world today, fossil fuels are still the primary source of energy used to generate electricity. Some of the common fossil fuels used include natural gas, petroleum, oil, and coal. Over time the cost of generating electricity using fossil fuels has increased as the supply has become more and more scarce. The cost of excavating these resources from the Earth has sharply inclined. These rising costs have negatively impacted consumers, and it has become apparent in the exorbitant utility bills that are due every month. While the initial cost of purchasing a solar power system is costly, you will save thousands if not more on utility expenses in the long run.

Essentially, all that you are paying for would be the equipment to construct your solar power system, and once it is working, you can begin enjoying the benefits of zero monthly utility costs.

3. Simple installation.

The cost of the equipment needed in constructing an off-grid solar power system has declined steadily over the years as more suppliers have entered the market and have developed newer technologies. However, companies who offer installation services have increased their charges, taking advantage of the growing demand of households ready to switch to solar power. Many people will end up hiring these expensive professionals because they perceive the process of installing a solar power system as being complicated or dangerous. This is the biggest misconception I find today concerning solar power systems. These systems are not the least dangerous to install, nor do they require a math or physics degree. Any homeowner with a set of tools can install a solar power system on their own, reducing the overall cost of investing in this technology.

4. Best alternative for remote living.

Access to electricity is a major stressor for people living in remote areas prone to receiving blackouts. The reason for a blackout is usually due to a lack of proper electrical infrastructure built to connect to the main electric grid. Investing in an off-grid solar power system will remove this burden forever. The independence from a utility company will save the household more money in the long run, but it will also provide the freedom to position the remote home in any open field, forest, or harbor. The remote home will become self-sufficient and complete control over the amount of power used within the home.

5. Creating a cleaner environment.

We have all heard of the vast amounts of pollution created as a result of burning fossil fuels. This pollution contributes to the prevalent social and environmental issue of global warming by releasing large amounts of carbon dioxide in the air, which traps heat in our atmosphere and influences climate change. Solar energy is a renewable energy resource, which means that it does not threaten our environment. We do not have to burn anything to generate electricity, which reduces our carbon footprint and preserves the air and nature around us. Solar power is not a threat to our ecosystem, instead, it allows us to use the natural energy coming from the sun that is free and available indefinitely.

Chapter 2: What to Consider Before You Start

Building your solar power system can be a very exciting project—it spells the beginning of freedom! However, this project will require your full commitment to making sure that there are no flaws in the system. These flaws can be prevented by the guidelines in the following chapters. However, before we can construct the system, there are a few considerations you must be aware of that will influence the size of your solar power system. In other words, each household will have a custom setup depending on a variety of factors. By exploring each of the factors which I will discuss below, you will effectively build a solar power system that meets all of your household and lifestyle needs.

Sun Exposure

Solar power systems rely on plenty of light for the household to receive sufficient energy. Even the smallest area of shade on the panels can reduce the quantity of solar energy that you can receive at any particular time. Therefore, you need to assess any obstructions that could be in the way of the solar panels receiving maximum light. If you have not yet built your remote home, you can position it in a location with minimal obstructions.

However, if you live in quite a shady location and you have already built your remote home (or permanent structure), the best solution would be to find an area in your yard that receives the most amount of sun during the day and mount your solar panels in the ground here. If you decide on the latter option, ensure that your solar panel array is mounted no more than 500 feet from your home. This is because the further the distance of the panels from the house, the larger the wire and conduit you will need to purchase (indeed, this implies more money being spent).

Amount of Power Needed

Not every household will utilize the same amount of electricity; some households do not own a single electrical appliance and others that cannot live without all of the essential electrical appliances. Therefore, each home will need to determine how much power they will need. When deciding this important factor, ensure that you have factored future needs and uses for electricity to avoid designing a system that you will outgrow. Some of the factors you may consider are whether or not you intend to have more children or live with friends or family, or perhaps you intend to start a small craft business in the future, and you will need to power some hand tools in your backyard. I would also recommend that you set realistic goals of how much power you can survive on a month-to-month basis, considering that an off-grid lifestyle is atypical to an urban city lifestyle.

Environmentally-conscious individuals can save the amount of energy that they consume by choosing the most energy-saving products that they can find at a local store. These energy-saving products may be up to 20 times more expensive than a normal energy-wasting product; however, in the long run, you will save a lot more energy. Let us make an example by comparing the amount of energy you spend when using a 10W LED light bulb in comparison to a 60W incandescent light bulb. On the surface, both light bulbs do a similar job; they offer the same amount of light, immediately switch on and off, and are both dimmable. Replacing three incandescent light bulbs with LED light bulbs would set you back about $30; however, if you used your LED light bulbs for four hours a day, you would be saving 600Wh a day.

60W incandescent light bulb X 3 bulbs X 4 hours = 720Wh

10W LED light bulb X 3 bulbs X 4 hours = 120Wh

Electricity saved: 600Wh

To effectively generate and store that wasted power in your solar battery system, you would need to spend between $1000 to $1500 more in purchasing extra panels and batteries. Therefore, the investment in purchasing energy-saving appliances is the most cost-effective solution in the long run.

Loads List

Once you have determined which energy-efficient appliances that you will use on a daily or weekly basis, you can proceed to create a loads list. A loads list is a list of all the electrical appliances that will be powered by your solar power system and measurement (in Watts) of the amount of power each device uses, how many hours during the day the device will be on and whether it has a start-up surge similar to a fridge or a well pump. At the end of your list you will determine the total amount of power that your appliances will use per hour every day. This list is crucial when you are ready to construct the off-grid solar power system because the total load will determine the number of solar panels required, the number of batteries, and the size of the inverter.

Calculating the Amount of Energy Your Home Will Require

Once you have assessed the location and considered how much power your tiny home will require, it is time to complete the calculations that will help you purchase the correct quantity of solar power components. These calculations are not complicated; all you need is basic math literacy and a willingness to learn. I will break down each calculation in the simplest form so you can follow this manual with ease. Let's begin.

Step 1: Analyze the load and give an estimate of the daily usage.

As discussed, determining how much power you need is a crucial step that cannot be overlooked or given a rough estimate. To avoid this step would be the same as driving toward a foreign destination without any map or coordinates. Important questions that you will need to answer at this stage are: What are you powering and how long do you need to power it? You also need to plan for worst-case scenarios, particularly weather conditions in your environment. For instance, during winter, the sky may be cloudy for weeks at a time, and you will need a solar power system that is large enough to provide power throughout these months.

Some people will try to do guesswork when it is time to decide on how much power an appliance requires. This is a dangerous exercise because your guess may be incorrect, leaving you with a system too small to power your entire home. Another mistake I see a lot is when people estimate the amount of power needed by considering the home's square footage. Once again this is a dangerous exercise because, within the same square footage, two cabins may be using different amounts of power (one cabin may be using an electric stove and a water heater while the other one is using a fridge, electric oven, air conditioning system, and an electric dishwasher).

Most electrical machines, appliances, and equipment will have a power label on them. Take time to look for this label because it contains all of the valuable information you will need to calculate how much solar power you need. Some labels will list the watts, amps, and others will add the voltage range. If you find that the voltage listed is DC instead of AC, you multiply the DC volts by the amps to receive a number in Watts. Other times the voltage listed will be a range; for instance, it might read 150 to 250 volts. In this case, you would use the voltage number that is generally used in your particular area.

Below are the first two formulas to calculate Watts and "daily W hours":

Volts X Amps = Watts

Watts X (hours the device is on per day) = Daily W hours

You will also find that some loads are not active; however, when they are on, they have a very high surge (this would be loads such as refrigerators, pumps, or heater fans). This makes it challenging to determine how many Watt hours the load uses in total throughout the day. If you find yourself in this predicament, I would suggest that you find the Energy Star label on the device or appliance; this label should provide you with the average annual kilowatt (kWh) used. Thereafter, take the kWh and divide it by the number of days in a year to get the kilowatt per day. Below is a photo of the power label on my washing machine with an annual kilowatt value. Let's use this formula below to help calculate the kilowatt hour per day:

KWh per year/365 days = kWh per day

192/365 = 0.53kWh per day

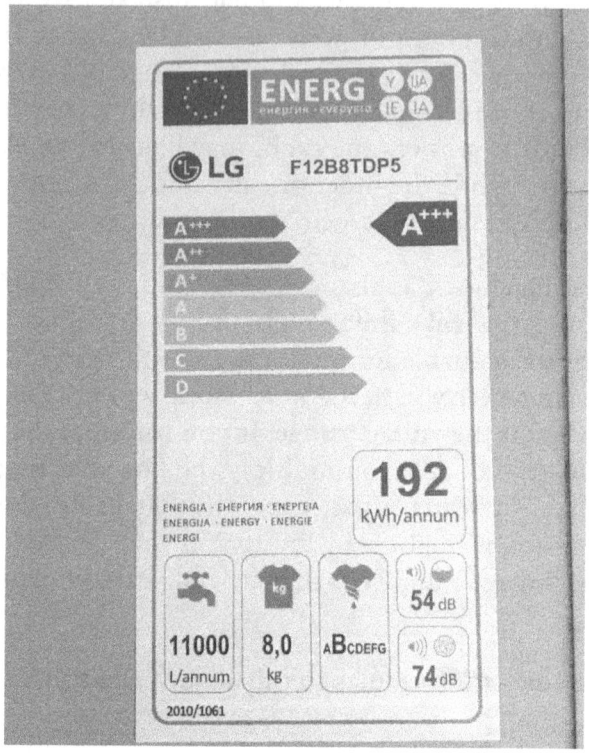

Step 2: Calculate the capacity of the battery bank.

This step involves calculating how big our battery bank should be. The first order of business is to select the voltage of the battery bank. Most off-grid battery banks are usually set at either 12V, 24V, or 48V. To determine the voltage, consider a few things. Firstly, the voltage size will be determined based on how large your load is. Generally speaking, you can always remember this rule: the higher the output wattage, the higher the DC input (voltage) needs to be. Alternatively, you can look at the size of the power inverter (which converts DC to AC) and determine the voltage using the size of the inverter. For instance, if you are using a 2000W inverter (perhaps the load is not too large), it will most likely use 24V, if you have purchased a 7000W inverter, it will most likely require 48V.

You need to also consider the type of battery that you will purchase to make the system as efficient as possible. Lead batteries are a popular choice for most off-grid solar power homes. When choosing a lead battery, you can either pick the flooded or sealed lead batteries. Flooded lead batteries are quite demanding in that they will need regular maintenance, involving adding water and assessing the gravity (sg) levels. On the other hand, sealed lead batteries will require less maintenance; they do not release a lot of gas and can be safely positioned in an unvented room (however not airtight). Flood lead batteries cost less than a sealed lead battery and have a longer life span, regardless of the amount of maintenance involved in taking good care of them.

The second type of battery available in the market is a saltwater battery, a fairly new non-toxic technology. One of the main features of this battery is that it can be discharged down to "100 percent empty" without harming the battery. This allows you to consume as much power as possible from the battery, thereby requiring fewer batteries for the battery bank. Indeed, this amazing benefit comes with a higher price tag than the lead batteries but you will reduce high costs by using fewer batteries in the battery bank.

Lastly, you can also consider using lithium batteries, which can be significantly discharged while also maintaining a long life span. The only limitation to using lithium batteries is that they are sensitive to being 100 percent emptied, and most of the time, you will have to replace them after one small incident. Therefore, when you use these kinds of batteries, you must create a battery management system that will monitor the proper charging and discharging of the battery. The cost of a lithium battery is higher than other available batteries in the market. Nonetheless, these prices should drop in the near future.

Now that you have determined how big the battery bank should be, you can start designing the battery bank. To illustrate this, assume that each battery is 12V and 200Ah (amp hour). To calculate the battery's power, multiply the two figures and get 2400Wh for one battery. You choose how to wire the batteries; either design them in a parallel or series. Remember that when you create a parallel circuit, it will increase the Ah but keep the voltage at the same level, and creating a series circuit will increase the voltage but keep the Ah at the same number.

Assuming that the total load that you would need to power is 4800Wh (information from our load list) and you wanted to create a parallel circuit, you would only need two batteries connected parallel to each other. However, if you need 9600Wh capacity, keep the design and add two more batteries, resulting in a four battery bank. Below is an illustration to show you the design of a parallel battery bank. Please notice how the amp hour changes and the voltage stays the same in both calculations:

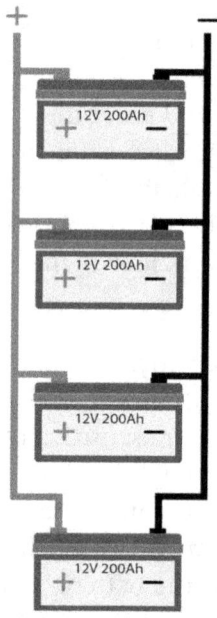

BATTERY: 12V x 200Ah = 2400Wh

(1 x 12V) x (2 x 200Ah) = 4800Wh = 12V, 400Ah

(1 x 12V) x (4 x 200Ah) = 9600Wh = 12V, 800Ah

However, you may not want to have so many strings in the circuit, so you need to tweak the design a bit. Instead of placing all of the batteries in parallel, wire them using parallel strings of two in series (instead of one long one). Below is the illustration of how the battery bank design would look like. Please notice how the voltage and the Ah increases as you combine both parallel and series circuits:

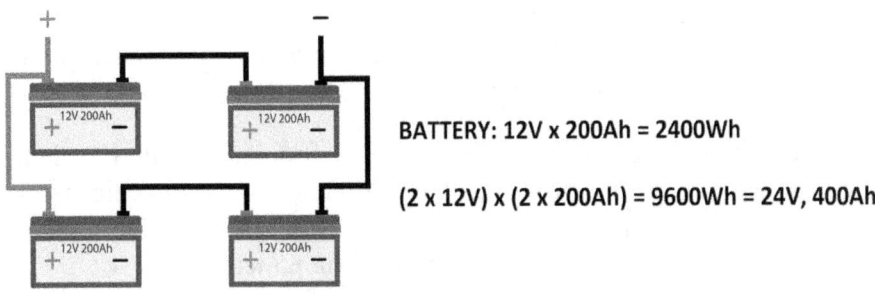

BATTERY: 12V x 200Ah = 2400Wh

(2 x 12V) x (2 x 200Ah) = 9600Wh = 24V, 400Ah

The day of autonomy refers to the amount of time (usually in days) that the solar power system can operate using battery power alone before you can use the generator. You can look at the amount of power you use daily or weekly to see how large the battery bank needs to last on days or weeks where there is no sunlight. Nevertheless, you need to remember that the more days you plan on using the battery alone, the larger and more expensive the battery bank will be. However, you do not want to underestimate the number of days either because you will unnecessarily strain the batteries (this will ultimately reduce the battery life span). After the days of autonomy are over, you can kick start the generator to charge the battery bank.

Step 3: Calculate the solar panel array wattage.

Make a few considerations when you decide on the number of solar panels needed for the solar power system. Firstly, you can look at an insolation map to show the peak sun hours in your geographical location. This can help determine whether you will manage with only a few solar panels because your area is always full of sun, or whether you need to invest in a large number of solar panels because you don't receive a lot of sun throughout the year. When choosing the number of panels for the solar array, always plan for the worst—in most cases, the worst-case scenario is always winter.

You should also plan for unexpected inefficiencies, such as the possibility of the solar panels being placed in an area with obstructions, the soiling of the panels, or having the voltage drop due to the wiring. When the solar panels do not work at their optimum level, you could lose as much as 1/3 of the solar power from the panels. For instance, in a 100W rated solar panel, you could potentially receive only 67W if the system had any inefficiencies, and the usable power would be significantly reduced.

Another factor that may contribute to the efficiency of the panels is the type of solar panels you decide to use. There are two basic types of PV solar panels in the market, namely monocrystalline and polycrystalline solar panels. Monocrystalline panels consist of uniform cells with a high-efficiency rating of up to 20 percent. These panels are readily available on the market, and they are the most reliable type of solar cells, having a reputation for lasting between 25 and 50 years.

Polycrystalline panels consist of non-uniform cells and, as such, have a lower efficiency rating of about 10 to 15 percent. The lower efficiency of each panel means that you need to install a lot more to get the right amount of power. Nevertheless, polycrystalline panels are a lot more affordable than monocrystalline panels, providing a sweet cost incentive for those seeking to reduce the total cost of installing their solar power system.

To calculate the minimum watts needed in the array of solar panels, you can perform a simple calculation: start with the daily watt hours and divide the number by your location's worst-case peak sun hours (during winter). Then take this new number and divide it by the system efficiency of 67% (calculated using the 67W from the example above). Below is an example of how your calculation should look:

Our daily watt hours: 2192Wh

Our peak sun hour: 2.8 (see insolation map for your area)

2192Wh/2.8/0.67 = 1168W

Once you have calculated how many watts of total panels the system requires, make sure that the number of panels you intend to purchase can be wired either using a parallel or series design. Furthermore, your responsibility is to ensure that the nominal voltage of the solar panels either matches or is higher than the voltage of the battery bank.

Step 4: Select a charge controller.

There are two types of charge controllers which you have the liberty to use as part of our solar power system. The two controller technologies are known as PWM and MPPT. The PWM charge controller works through making a connection from the solar array directly to the battery bank. When there is a continuous active charge from the solar array to the battery bank, the array's output voltage is pulled down or lowered to the battery voltage. The more the battery charges, the higher the voltage of the battery increases, thereby increasing the total voltage output of the solar panel simultaneously. Make sure to match the nominal voltage of the solar panels with the voltage from the battery bank. A 12V panel is required to charge a 12V battery, and a 48V panel must be used to charge a 48V battery, and so forth.

MPPT controllers on the other hand will measure the Vmp (voltage at maximum power) of the solar array and convert the high PV voltage into a lower battery voltage. When this voltage is lowered to match the battery bank voltage, the current is effectively raised, and you end up receiving far more available power from the solar panels. Due to this technology, MPPT charge controllers are usually more expensive than PWM controllers; however, they are significantly more efficient than the latter.

Step 5: Specify the size of the inverter

This is the final step of the design and calculations process. In this step, you will need to select a size for the inverter you desire to place in the off-grid solar power system. At this point, you have completed most of the major decisions; you have successfully determined the house load, figured out the number of batteries and solar panels required, and have chosen your preferred charge controller. When deciding on the size of an inverter, remember the function that it is created to serve. The inverter's purpose is to convert the direct current (DC) from the battery bank into an alternative current (AC) that can be used to power the house load.

The DC voltage must be changed into a sine wave curve that travels above and below 0 volts to do this. When inverters were first invented, the most appropriate way to do this was to make the voltage rise straight up and come straight down, creating what looked like a chopped signal. This chopped up and down signal is referred to as a Modified Sine Wave. One of the benefits of using a Modified Sine Wave inverter is that it is useful when building simple systems or powering old electrical appliances built using older technologies. This type of inverter would not be suitable for powering electronic equipment, audio, rechargeable batteries, or digital clocks.

Over the years, inventors have actively sought to improve upon the Modified Sine Wave technology and, as such, have added multiple steps, attempting to come as close to a pure sine wave curve as possible. The Pure Sine Wave produces a more smooth and consistent voltage signal, rising up and down effortlessly. This smooth signal allows the house loads to receive the best possible current, making it more appropriate for off-grid remote homes. Pure Sine Wave inverters are more costly than Modified Sine Wave inverters; however, the benefits are tremendous, especially in powering inductive loads, which pull a large amount of electrical current when they are first turned on.

Now that the charge controller has charged your battery bank, the off-grid solar inverter can convert the 12VDC, 24VDC or 48VDC battery bank into an AC voltage. The AC output will largely depend on your specific needs; however, in North America, you can make use of a 120V single phase, 240V split phase, 480V 3-phase, and so forth. Once again, it will depend on how you wire the output of your inverter, the type of inverter that you will purchase, and determining what your house loads require.

It is also good to remember that an off-grid inverter is not able to feed excess power back to the main utility grid; however, it can connect to the main utility grid and use it as a battery charger. For instance, those who live in a houseboat or RV can connect to shore power and use the main grid to charge their battery bank when their solar panels do not provide sufficient power. However, the AC connection will remain one-directional, meaning that it will only take from the grid and not give any power back.

Lastly, you will need to insert breakers into our solar power system to protect from excessive currents and short circuits. There are five main areas where you can place breakers in the system: between the solar panels and the charge controller, in the solar combiner box as well as in the DC load center, between the charge controller and the battery bank, between the battery bank and the inverter, and lastly between the inverter and the AC loads.

Materials and Tools for Building an Off-Grid Solar Power System

We have discussed the various components that make up an off-grid solar power system; however, there are a few supporting materials and tools that you must have when installing the system to perform successfully. These materials and tools can be purchased from a local DIY store or Home Depot and stored in your home for future solar projects.

Solar systems mostly require standard household tools such as a screwdriver, hammer rails, ladders, etc. Nonetheless, there are a few custom tools that you will need to purchase, which are designed to help you install the solar system. Some of these tools are utilized for simple applications, and others for more complex operations. For those who are looking forward to installing a home's off-grid solar power system without professional help (effectively making it a DIY project), you need to make sure you have purchased tools for physical application, such as things to install racking. Furthermore, some states require an electrician to visit the site and check the connections before switching on the system. Contact the local authorities and inquire about any procedures that you need to follow before and after successful installations.

1. Safety comes first.

I have put safety equipment at the top of the list because you cannot perform any task without protecting yourself first. Safety equipment and accessories such as ropes, ladders, harnesses, safety goggles, gloves, safety suits, earplugs, helmets, and safety boots. I recommend you purchase as much safety equipment as you can because it will reduce any risks of injury when you are installing your system.

2. Site assessment.

You will need materials and tools to perform a site assessment. Every successful installation has to begin with establishing the layout of the solar array. Therefore, you will need a fairly large tape measure (consider using two tape measures one being a 25-foot and the other being a 100-foot), white chalk or string line to demarcate the area where to position the array, and a solar pathfinder which will show the most efficient solar array location and position on your plot. Don't feel you have to rush the planning process because any mistake in positioning the solar panel array can cost a lot of time and money in the future.

If you have chosen to mount the panels on a rooftop, you will need a suitable sized ladder to reach the area to place the panels. When the layout is on a pitched roof, find the location of the roof rafters or trusses. Do this by tapping on the roof with a hammer to audibly sense the location of the roof's structural support. Once you have found a rafter, mark the location with a pencil and measure about 18 to 24 inches, reaching the next rafter and placing a pencil mark there. Once all of the rafters (top and bottom) are marked you can start drawing lines to mark the rafters lengthwise and follow by marking the locations of the stand-offs as well as the rails to complete the pitched roof layout.

Planning installation for flat roofs is far less intricate because the measurements only require you to decide whether the solar panel array will fit perfectly on the roof's surface area. However, if you decide to mount the solar array on the ground, locate the support posts for the racking. This can be done by contacting the local utility company to find and mark the underground lines (this service will be free of charge and must be done before you can proceed with the layout). A solar pathfinder is a useful instrument for this ground-mounted solar layout because it will show you the best position for the structure (azimuth angle).

3. Structural Support Materials

Depending on the method desired to mount the solar array, you will need different structural support. For instance, if you are mounting the panels on a pitched roof, you will need to have a cordless drill as well as an impact drive, drill bits, a utility knife, sockets, and a caulking gun. Ensure to first drill a pilot hole so you don't split the rafters; a pilot hole should also make installing a stand-off anchor bolt a lot more convenient using an impact driver. The utility knife will come in handy by trimming off any roofing material to install the flashing (inserted around joints or cracks on a roof to provide a water-tight seal).

Flat roof structural support is a lot simpler. Most of the time, a ballasted racking system will provide the support for the panel array. You might need to use a broom to remove any dust or dirt on the flat room, especially within the installation area. The only difficult part of this installation will be carrying the heavy ballast weight onto the roof. In extreme cases, you can rent a forklift or a crane to lift the weights. Ground-mounted solar arrays require holes in the ground to anchor the posts in concrete. Digging holes is difficult and time-consuming, so schedule enough time and human resources to effectively complete this task. Ensure that the posts are planted straight in the ground by using string lines as markers.

4. Racking

Once the structural support is installed, you can proceed to install the racking. The racking will help fix the panels securely on any surface. The method of inserting the racking is similar for every mounting type and usually you will only need to use one socket size for the whole racking installation. Use a torque wrench to insert all the necessary bolts and a cordless reciprocating saw to install the rails on one side of the array and only trim the rails according to the perfect fit once all of the panels have been installed seamlessly.

5. Grounding

All PV solar power systems require the equipment to be grounded so that they will not move or shift from their location. Ground the equipment with wire cutters, using a flat-head screwdriver, using a drill for installing lugs, and a flat-head screwdriver for tightening the installed ground lugs (inserted on the back of roof-mounted systems).

6. Wiring

Even though connecting the panels in series will not require any tools, you will still need the tools for taking panels apart. Another great use for tools is creating a home-run wire that utilizes custom crimping tools to securely establish a connector on the appropriate wire matching the panel connector. Once you have successfully installed a positive and negative home-run for each string, you need to inspect the voltage by using a multimeter. When you inspect the voltage after each string is in place, confirm you have properly installed the wires and reduce the risk of experiencing issues later on in installation.

Calculation of the wire section (12V DC)

Choosing the right wire sizes in your PV system is important for both performance and safety reasons. If the wires are undersized, there will be a significant voltage drop in the wires resulting in excess power loss. In addition, if the wires are undersized, there is a risk that the wires may heat up to the point in which a fire may result.

An electrical wire carries current much like a water hose carries water. The larger the diameter of the water hose, the less resistance there is to water flow. Moreover, even with a large diameter hose, shorter hoses have better flow than longer hoses. Longer hoses have more resistance than shorter ones of the same diameter. Electrical wires behave the same manner. If your electrical wires (the copper gauge) are not large enough or if the cable is longer than needed, then the resistance is higher resulting in less watts going to either your battery bank.

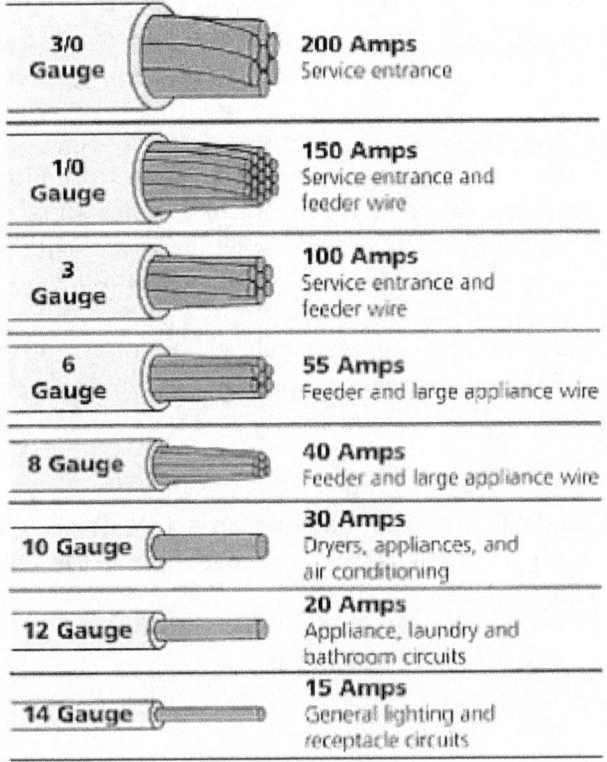

Commercial solar PV panels over 50 watts or so use 10 gauge (AWG) wires. This allows up to 30 amps of current to flow from a single panel. If multiple panels are combined in parallel, then a three to eight AWG "combiner" wire set is generally needed to safely transfer the power to a charge controller.

The wires from the charge controller to the battery bank can generally be the same or larger gauge than the main set from the PV panels.

The wires between batteries in a battery bank tend to be the largest in the system since they are used in conjunction with a power inverter that can at times demand more current than that the PV system can supply on its own. These same wires will also have to carry current used simultaneously for charging and for power inversion.

A typical battery bank wire size is 1/0 or "one-ought."

It is very important to match the gauge and the wire lengths when combining batteries in a battery bank. If this is not done, then the battery bank's life can be shortened and certain safety issues can result.

Usually the longest wire run is from the PV array to the location where the charge controller or GTI is located. Since all of the combined PV power flows through this wire set, we really need to choose it correctly to maximize performance and to assure safety.

The general rule-of-thumb is to stay below 2% Voltage drop on this run. Using the known resistance of the various wire gauges, it is possible to calculate the maximum length for a wire-pair for each wire gauge size.

Here is what that calculation looks like for a 12V PV system. You can double the length for a 24V system, or quadruple it for a 48V system.

Example: Let's take a 450-watt 12V system. At the Vmp of 18V, the maximum current is 450/18 = 25 amps.

	2% Voltage Drop Chart					
AWG =	14	12	10	8	6	4
Capacity(AMPS)	15	20	30	40	55	70
ARRARY AMPS	FEET ONE WAY FOR A PAIR OF WIRES					
1	45	70	115	180	290	456
2	22.5	35	57.5	90	145	228
4	10	17.5	27.5	45	72.5	114
6	7.5	12	17.5	30	47.5	75
8	5.5	8.5	11.5	22.5	35.5	57
10	4.5	7	9.5	18	28.5	45.5
15	3	4.5	7	12	19	30
20	2	3.5	5.5	9	14.5	22.5
25	1.8	2.8	4.5	7	11.5	18
30	1.5	2.4	3.5	6	9.5	15
40			2.8	4.5	7	11.5
50			2.3	3.6	5.5	9
100					2.9	4.6

Looking at the wire capacity row, 10 AWG is the smallest gauge wire that can safety be used. It is rated at 30 amps, higher than the required 25 amps.

Next, we look at the Array amps column, select row "25" and you can see that a 10 AWG wire pair only supports a cable length of 4.5 feet. Going up to 4 AWG supports up to 18 feet to stay within the 2% loss criteria.

What this example illustrates is that we need to greatly appreciate the issue of cable length and its effect on losses. Many people have long cable runs and don't realize the impact this has on performance. Sometimes we have to tolerate perhaps a 4% loss rather than 2%, allowing us to double the length values shown in the table. Another option is to operate at a higher voltage, such as 24V. This reduces the amps which reduces the wire losses.

Calculation of the wire section (120V/220V CA)

To calculate the AC current (wires after the inverter) that could pass through, we need to do a simple calculation using Ohm's law for AC. Knowing the power of the inverter we have installed, assuming it is 1500W at 120V AC, we do:

$$I = W/(V \times \cos\phi)$$

Where **W** is the power (1500W), **V** is the voltage (120V AC) and **cosϕ** is the power factor which is assumed to be 0.9.

$$I = 1500/(120 \times 0.9) = \mathbf{13{,}88A}$$

So our current is 13.88 amps. Now we can go to look for the closest value in the table above and do the same way done with the calculation of the wire in DC.

Choice of Battery

Choosing the right batteries for your solar systems is so important because different batteries have varying levels of energy efficiency, storage capabilities, and cost effectiveness. There are some challenges related to choosing the right solar battery, such as battery's capacity & power ratings, depth of discharge (DoD), performance, battery life & warranty, charging time, and your gudget.

Capacity & power

Capacity is the total amount of electricity that a solar battery can store, measured in kilowatt-hours (kWh). Most home solar batteries are designed to be "stackable," which means that you can include multiple batteries with your solar-plus-storage system to get extra capacity.

While capacity tells you how big your battery is, it doesn't tell you how much electricity a battery can provide at a given moment. To get the full picture, you also need to consider the battery's power rating. In the context of solar batteries, a power rating is the amount of electricity that a battery can deliver at one time. It is measured in kilowatts (kW).

Battery capacity is typically measured in unit of Amp-Hour (Ah), it states the amount of energy that a battery can hold under specified conditions. For instance, a 12 volt 18 Ah battery can delivery one hour of 18 ampere of current at 12 volt.

Keep in mind that if you get stackable batteries, you can use multiple batteries at once. One thing to think about is that a battery with a low capacity and high power rating can run your whole home, but for just a few hours, whereas one with a high capacity and low power rating can only power a few important appliances, but for a long time. Generally speaking, the higher the capacity and the higher the power rating, the better off you are.

Depth of discharge (DoD)

Most solar batteries need to retain some charge at all times due to their chemical composition. If you use 100 percent of a battery's charge, its useful life will be significantly shortened.

The depth of discharge (DoD) of a battery refers to the amount of a battery's capacity that has been used. Most manufacturers will specify a maximum DoD for optimal performance. For example, if a 10 kWh battery has a DoD of 80 percent, you shouldn't use more than 8 kWh of the battery before recharging it. Generally speaking, a higher DoD means you will be able to utilize more of your battery's capacity.

Performance

A battery's performance represents the amount of energy that can be used as a percentage of the amount of energy that it took to store it. For example, if you feed five kWh of electricity into your battery and can only get four kWh of useful electricity back, the battery has 80 percent efficiency (4 kWh / 5 kWh = 80%). Generally speaking, a higher performance means you will get more economic value out of your battery.

Battery life & warranty

For most uses of home energy storage, your battery will "cycle" (charge and drain) daily. The battery's ability to hold a charge will gradually decrease the more you use it. In this way, solar batteries are like the battery in your cell phone – you charge your phone each night to use it during the day, and as your phone gets older you'll start to notice that the battery isn't holding as much of a charge as it did when it was new. For example, a battery might be warrantied for 4,000 cycles or 10 years at 80 percent of its original capacity. This means that at the end of the warranty, the battery will have lost no more than 30 percent of its original ability to store energy.

Your solar battery will have a warranty that guarantees a certain number of cycles and/or years of useful life. Because battery performance naturally degrades over time, most manufacturers will also guarantee that the battery keeps a certain amount of its capacity over the course of the warranty.

The general range for a solar battery's useful lifespan is between 5 and 15 years. If you install a solar battery today, you will likely need to replace it at least once to match the 25 to 30 year lifespan of your PV system. However, just as the lifespan of solar panels has increased significantly in the past decade, it is expected that solar batteries will follow suit as the market for energy storage solutions grows.

Proper maintenance can also have a significant effect on your solar battery's lifespan. Solar batteries are significantly impacted by temperature, so protecting your battery from freezing or sweltering temperatures can increase its useful life. When a PV battery drops below 30° F, it will require more voltage to reach maximum charge; when that same battery rises above the 90° F threshold, it will become overheated and require a reduction in charge. To solve this problem, many leading battery manufacturers, like Tesla, provide temperature moderation as a feature. However, if the battery that you buy does not, you will need to consider other solutions like earth-sheltered enclosures. Quality maintenance efforts can definitely impact how long your solar battery will last.

What are the best batteries for solar?

Batteries used in home energy storage typically are made with one of five chemical compositions: lead acid, lithium ion, AGM, gel batteries, and Saltwater. In most cases, lithium ion batteries are the best option for a solar panel system, though other battery types can be more affordable.

1) Lead Acid - Lead acid batteries are a tested technology that has been used in off-grid energy systems for decades. While they have a relatively short life and lower DoD than other battery types, they are also one of the least expensive options currently on the market in the home energy storage sector. For homeowners who want to go off the grid and need to install lots of energy storage, lead acid can be a good option. When it comes to drawbacks, flood lead acid batteries tend to be very large and take up a lot of space, plus they also tend to suffer from corrosion. This type of battery needs a fair bit of maintenance and monitoring, as they need to be kept in a well-ventilated environment, and they need to be kept upright to prevent leaking. If you don't mind some regular maintenance and monitoring, this is one of the best options to go with, but if you don't want to engage in regular monitoring, the AGM battery may be a better choice for you.

2) Lithium Ion - The majority of new home energy storage technologies, such as the , use some form of lithium ion chemical composition. Lithium ion batteries are lighter and more compact than lead acid batteries. They also have a higher DoD and longer lifespan when compared to lead acid batteries. However, lithium ion batteries are more expensive than their lead acid counterparts.
They have great efficiency and really long lifespans and are also very safe and stable; they barely require any maintenance, and don't require any attention for at least 15 years after installation.
They also do not lose as much capacity as other batteries when they are idle. All in all, for more or less every single reason, lithium ion batteries are the best, but they are also by far the most expensive ones to go with.

3) AGM (Sealed Lead Acid) - This is a type of lead acid battery, but one where the electrolytes are absorbed into a glass matt. The plates in an AGM battery can be flat or they can be wound into a spiral.
One of the advantages is that this type has a lower internal resistance than flood lead batteries, plus they are better at handling temperatures, and they do not discharge as slowly as some other battery types.
Other advantages include that AGM batteries hold static charges for a long time, they are not inclined to heat up, they can handle the cold quite well, they are lightweight, and non-hazardous too. They also do not require as much maintenance as flood lead batteries. With that being said, they come with quite a hefty price tag.

4) Gel Batteries - Most people probably won't go with gel batteries, as they have more drawbacks than most of the other options, and their advantages are fairly minor when compared to other battery types. Gel batteries used fumed silica to thicken the electrolyte, which makes these cells sturdier. What is also convenient is that the viscous nature of the electrolyte prevents leaking when damage occurs.

Other advantages of the gel battery are that it does not require much maintenance at all, and they are quite durable, as well as resistant to shock and vibration. Moreover, they are great both for extreme heat and cold, plus they have excellent life cycles.

With that being said, gel batteries have narrow charging profiles and it is not hard to damage them through improper or over charging. These also don't have the greatest amp-hour capacity, but they still cost quite a bit.

5) Saltwater - A newcomer in the home energy storage industry is the saltwater battery. Unlike other home energy storage options, saltwater batteries don't contain heavy metals, relying instead on saltwater electrolytes. While batteries that use heavy metals, including lead acid and lithium ion batteries, need to be disposed of with special processes, a saltwater battery can be easily recycled. However, as a new technology, saltwater batteries are relatively untested, and the one company that makes solar batteries for home use (Aquion) filed for bankruptcy in 2017.

BATTERY	COST	LIFESPAN	DoD
Lead Acid	$	▲	▮
Lithium	$ $ $	▲ ▲ ▲	▮ ▮
AGM	$ $	▲ ▲	▮ ▮
Gel Batteries	$ $ $	▲ ▲ ▲	▮ ▮
Saltwater	$ $	▲ ▲	▮ ▮ ▮

Charging Time

If you are worried about how much sun and time it takes to charge the solar battery in question, you are probably best off going with gel, as they charge really fast. Lithium-ion batteries also do not take long to charge, usually around 3 hours or so, maybe a bit longer.

Lead acid and AGM batteries can take up to a whole day to charge, if not longer. Of course, the faster the battery charges, the better off you are because it doesn't take as much light or time for charging, and you can get to using that solar power much faster too.

Your Budget

The upfront cost for the solar battery itself is something else you should keep in mind. One thing we want to note here is an old adage – you get what you pay for. Generally speaking, the more expensive the solar battery, the better it will perform and the longer it will last.

Something like a lithium-ion battery might cost many times the amount of an AGM or lead acid battery, but it also requires much less maintenance, is more resilient, easier to install, and features better overall performance in pretty much every way.

Calculating Costs

Even though solar power is the most affordable option, the immediate investment will be steep in the long run. However, this investment is a once-off commitment that you will begin to profit from as soon as your system has been turned on. Once you are officially off-grid, your monthly expenses will drop exponentially, and you will know how it feels to be completely self-sufficient.

When calculating the total cost involved in going off-grid, you can immediately deduct around $5000, which would have gone toward paying an installation company to install the system. The remainder of the costs will be different for each household because not everyone has the same power needs. In other words, the cost varies depending on your particular lifestyle. The less electricity you need to power your remote home, the less expensive the solar power system will be. Nonetheless, to give you a general picture, a solar power system would cost an average American household that uses a minimum of 7kW of electricity every day between $30,000 and $60,000 to install (consisting of six solar panels and three batteries). However, you can also break down the cost to purchase each component of an off-grid solar power system.

Solar panels will be the most expensive component of your off-grid solar power system. This is because, without the solar panels, you would not have a device to harness solar energy and begin producing electricity. The number of solar panels, the number of solar cells, and the type of solar cells will vary depending on your unique needs. However, roughly you can expect to pay around $14,000 for solar panels alone.

Solar batteries are important because they are what allow for a home to go completely off the grid. All of the excess energy that your home cannot use on any particular day will be stored in batteries. This allows you to have electricity at night or when the weather conditions are not favorable. Once again, the size of a battery bank, the type of battery used, and the batteries' power will vary between households. However, you can expect to pay between $5,000 to $8,000 for each battery (typically, you will need between 2 to 4).

Solar inverters are incredibly useful to a solar power system because they convert solar energy into electricity that you can use to power your home. The size, type, and power of the inverter will depend on the amount of electricity, the number of solar panels, and the quantity of batteries the system has. The cost of an inverter tends to start around $3,000; however, it will increase depending on requirements.

A solar charge controller transports solar energy from the PV solar array to system loads and the battery bank. Each battery will, therefore, need to have its own controller, and the cost for one of these devices starts at around $550 each.

Chapter 3: Building Your Off-Grid Solar Power System

Now that I have shown you all of the necessary calculations and preparations that you need to make before constructing your off-grid solar power system, you can now proceed to build the system. Individual remote homes will all require a different amount of solar panels and batteries, depending on power requirements. Therefore, whether you live in a cabin, RV, or boat, you will have the same opportunity to receive as much electricity as your home can carry. I would suggest that you only purchase solar power system components and equipment after doing the necessary calculations. This will allow you to base your solar requirements on your calculations and not on the products on special.

DIY: Solar Power for Tiny Homes and Cabins

In their effort to minimize their dependence on materialistic possessions and free themselves from relying on the state for power, many homeowners will choose to live off-grid in remote homes located on the outskirts of the city. These smaller remote homes may be tiny versions of normal houses or cabins. Despite their size, these remote homes still require electricity for lighting, heating, bathing, and cooking. If you are an individual who is planning on building a tiny self-sufficient home or want to disconnect your home from your local utility grid, building your very own off-grid solar power system is the best solution for you. It will allow you to generate clean, renewable energy and power your home with as much electricity as you require (without paying costly bills each year to keep your lights on).

Some of the benefits that you will receive by building your own off-grid solar power system are similar to the benefits that many enjoy by shifting toward solar energy. In the long run, you will save money, produce clean energy that does not pollute the environment, and enjoy full control over your home's energy generation. In the future, when you decide to sell your tiny home, the presence of a solar power system will significantly increase the value of your property, allowing you to sell it for even more money.

Key Questions to Ask Before Building

When designing an off-grid solar power system for your tiny house or cabin, there are some important questions which you need to answer before making any major decisions. These answers will help you curate your system depending on your specific electrical requirements for your home.

How much electricity do you intend on generating?

Typically, tiny homes will use much less electricity than the amount used in larger homes. Nonetheless, each tiny home's load will be different due to how each household decides to live, the appliances that are used, and the number of people living within the home. To save a lot of money, you will want to generate as much electricity as possible. This means that it is good to have as many solar panels as you can afford.

Are you installing a roof-mounted system or a ground-mounted system?

Decide where the system will be positioned before anything else. If you intend to mount the solar panels on the roof, it is important to check whether the roof receives an ample amount of sunlight. Also, if the roof is strong enough to hold and handle the weight of the solar panels without compromising its integrity, alternatively, you can choose to mount your solar panels on the ground. These ground-mounted systems will require extra racks as well as extra mounting equipment. Lastly, make sure your panels are slightly tilted to capture maximum sunlight.

How many batteries will you need?

Since you are using an off-grid solar power system, you will need to purchase a certain number of batteries to store excess solar power. The cells will form an indispensable part of the system because they will allow plenty of electricity during the night or when the weather conditions are not favorable for powering solar panels. If you are going to purchase batteries, consider storing them indoors as they are sensitive to extreme weather fluctuations and can wear and degrade fairly easily when having to adjust between freezing and hot temperature regularly. Many suppliers will sell complete solar kits that include batteries that you can install in your home (this will remove the hassle from having to decide on the number of cells you have to buy).

Calculations for Your Off-Grid Tiny Home or Cabin

A cabin that only uses a propane stove, heater, a few lights, and a ceiling fan will have lower electricity requirements than a cabin that uses a washing machine, running fridge, electric stove and other energy-consuming appliances. In essence, determining your house load will be the first calculation you would need to make to determine the size of your solar power system. Let's make a few calculations and determine how many panels an average tiny home would require:

Hours of direct sunlight received per day: 5 hours

The average daily energy consumption: 100kWh per month

The wattage of each solar panel: 290W

Using the figures above, the average solar power panel will produce 290W, multiplied by the number of hours of direct sunlight will give an amount just under 1,500kWh or 45kWh per month. This means that one solar panel has the capacity of 45kWh per month. To meet the 100kWh per month, you will need to add a few more panels:

1 standard solar panel = 45KWh

2 standard solar panels = 90kWh

3 standard solar panels = 135kWh

4 standard solar panels = 180kWh

Therefore, a tiny home would need around 2 to 3 solar panels to meet power needs. However, to create extra room in the system for unexpected power usage it is recommended you purchase 4 standard solar panels for this type of home.

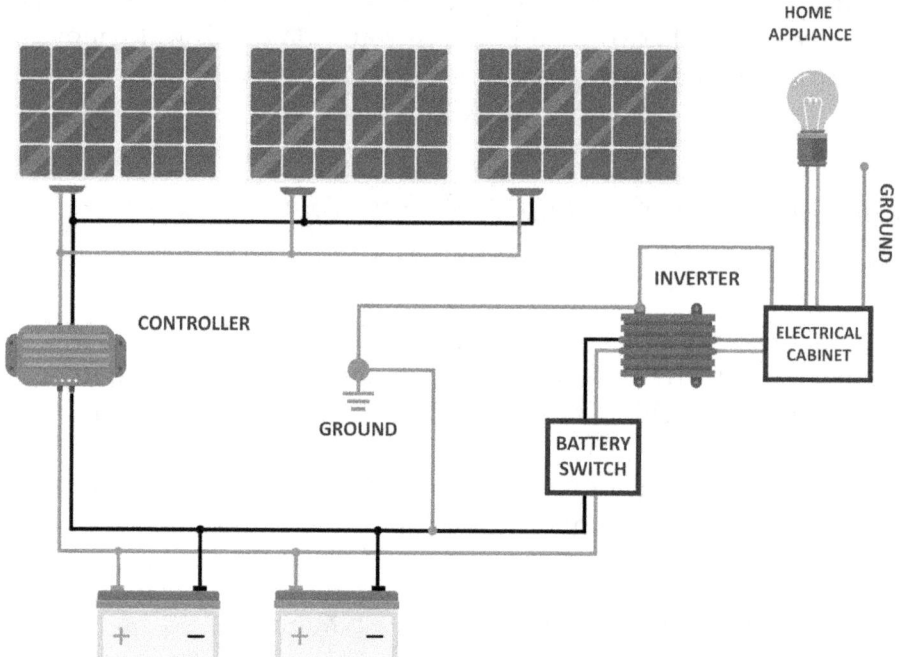

DIY: Solar Power for RV's and Boats

Small boat and RV off-grid solar power systems are identical in how they are structured but vary in size (depending on the power requirements). For instance, one home may only require a 5W panel, which keeps the solar batteries charged between a few road trips or sailing trips a year. Otherwise, when homeowners live in their small boat or RV full time, they may require 900W of solar panels, a large battery bank, and an inverter. I also find that generators are a common feature on most off-grid RV and sailboat solar systems.

Most RV and boat light bulbs and small accessories run on 12 volts DC; these lights are easily powered by solar panels and batteries. To reduce the amount of power needed in a solar power system, switch to more efficient electrical appliances. For instance, you can significantly reduce the number of solar panels required in your system by using DC fluorescents (a great product would be the Thin-Lite 12V fluorescent light bulb). These bulbs will give you the same power as a normal incandescent light bulb but only use around a third of the amount of power that incandescent light bulbs consume.

However, I would caution you in purchasing only 12-volt appliances because they tend to be of lesser quality (and more expensive) than standard household appliances. I also find that most RV's will come with some sort of inverter which ranges from 250W for operating small electrical appliances and reaching 4,000W pure sine wave, including a built-in battery charger.

Key Considerations Before Building in Your RV or Boat

The number of panels required on each RV or boat will also vary depending on power use. For instance, if you only plan on powering the lights, a small TV, and the common gadgets that come installed in a mobile home, expect to use around 80W to 130W of panels as well as a deep-cycle solar battery. However, if you will need an inverter to power a microwave, vacuum, coffee machine and other AC appliances, you would generally need about 200W to 450W of solar panels.

Batteries

Each home will have different battery needs; however, small RV's and campers will need a minimum of around 200 to 225 amp hour capacity. However, also consider the amount of space you have to store a battery bank, limiting the amount of batteries you can purchase. For instance, you may need 800 amp hours of capacity but only have room to store half of that—this would limit the amount of power that you can use in your home. When you are connecting your batteries, it is important to remember not to use old batteries with new batteries in one set or to combine different types of batteries in one set because it could cause problems in your system.

To calculate how much battery power you will need (in amp hours) you will need to work out how many watts you will require every day. One can assume that your battery bank should at least supply twice as many watts as you use each day, in order for you to never run out of electricity. For instance, a 12-volt battery that can supply 100 AH will give 1200 watts; however, don't aim to use the full 1200 watts because that would kill the battery. Instead, leave 50% power in each battery or use 600 watts at most.

Inverters

Some prefer to use a small inverter that is reasonably priced and can handle light loads such as powering laptops and other electronic devices. However, others will require a large inverter that can run coffee machines or microwaves. These larger inverters can cost between $500 and $1,100 and supply 800 to 1200 watts of power. Most smaller RV's or houseboats will use inverters that can provide around 1000 to 1500 watts of power.

Hybrid Systems

A hybrid system refers to the use of solar power and a generator in producing a sufficient amount of electricity in the home. I find that many people will use solar for powering small loads such as lights or electronic devices (and use a small inverter carrying a 250 to 800 watt capacity for running the television or computer). Nonetheless, there are times when you need to power heavier loads, and times like this are when you need a generator.

To save on cost, don't run the generator all day. Doing this would be expensive (in terms of the fuel needed to run it for 12+ hours), but it would also infuriate your neighbors with the noise. Therefore, only power the generator up to 2 hours a week and you can use this time to charge a battery bank. Boats specifically use a hybrid system which consists of solar panels and a wind generator. When living on an RV or houseboat, you can reduce dependence on a generator by making sure the panels provide more than 130 watts of power (this will reduce the need for a generator by 60-80%).

Battery Chargers (Converters)

Some RV's come with built-in battery chargers which only charge batteries at the same rate that 1 to 2 panels would (approximately providing 3 to 10 amps). Even though modern RV's have more sophisticated battery chargers, the average one you will find has poor regulation and charging characteristics. Furthermore, many manufacturers will refuse to replace batteries that have been charged through using a converter because these converters are known to burn the batteries if left charged for too long. Therefore, I recommend avoiding using a converter altogether and simply relying on purchasing more solar panels, better quality and type of batteries, and an inverter with a great 3-stage charger.

Mounting a Solar Array

For most boats and RV's, I would recommend a flat or tilted mounting system. There are many benefits of purchasing tilting mounts; one of them is they offer 10 to 20% more power production under normal weather conditions. However, tilting your solar array on an RV rooftop can be a hassle for those always traveling because it requires the RV to be parked for the vehicle to face due south and take advantage of the sunlight. Furthermore, I would not recommend the tilting mounts be left on the roof when the vehicle is moving because it could damage the supporting structure. Therefore, you should consider installing your solar array flat on the RV roof to limit shifting when the vehicle is moving and to take full advantage of the sun at all times.

Calculations for an Off-Grid RV or Boat

When deciding on building an off-grid solar power system for an RV or houseboat, you cannot ignore the cost implications that may be involved in owning one of these systems. Therefore, you will need to do some calculations to determine whether one of these systems will be feasible to own. Below are simple counts of a standard medium power usage RV or boat system:

- 1 or 2 X 80 to 150-watt solar panels.
- 1 X panel mount, which you will install on the roof of your vehicle or boat.
- 1 X solar charge controller.
- 4 X 220 amp hour 6-volt deep cycle batteries (a battery that can power a golf-cart).
- 1 X 600 to 1,500-watt solar inverter.

The total estimate for this kind of system varies from $400 for smaller systems, reaching up to $4,000 for larger systems (with some of these large systems including a sine wave inverter). With the example used above, the total energy output per day would come to 1600 watt hours. This would be enough power to effectively run the lighting system as well as the appliances in your mobile home with no issues.

Schematic for boat with isolation transformer and connection to the quay.

Isolation transformers are the only way to completely eliminate the risk of galvanic corrosion between the ships hull and other metal objects (other ships, pilings etc.) and ensure safety for swimmers.

The problem is caused by an earth loop forming and current flowing through the earth wire on the shore supply. The problems occurs in all boats, although steel & aluminium boats are particularly susceptible. It is important to get advice from a suitably qualified marine surveyor or corrosion specialist before installing shorepower.

The isolation transformer eliminates any electrical continuity between shorepower and the ship. The shorepower is fed to the primary side of the transformer and the ship is connected to the output side (secondary).

An isolation transformer takes your marina's often wild and unpredictable 120VAC shorepower and converts it to pure clean power. And by creating an onboard power source, it greatly enhances the safety of those on your boat or swimming nearby.

Most of us know how important the green ground shorepower wire is. It carries fault current (electricity that's going somewhere it's not supposed to, like when shorepower shorts against a metal case onboard) back to shore where it can't hurt anyone.

But marina shorepower systems may be less than reliable. Due to long-term corrosion or improper installation, the ground wires are sometimes not properly connected, meaning you (and nearby swimmers) are not protected from a fault if the AC shorepower shorts into the DC system. This could happen because of a problem in any AC/DC appliance, such as a battery charger. If that happens, any fault current is going to follow a path all through the boat's DC ground and bonding system, which is connected to the engine and underwater fittings, such as thru-hulls and prop shafts. Because leaking current always searches for a way back to its source (in this case, the marina's shorepower system ashore), leaking current will exit the boat and head toward shore. If a swimmer passes through the current, they will be electrocuted and may be killed.

The useful thing of the isolation transformer is that because it's taken over duties as the boat's power source, any leaking current will simply return to the transformer on the boat, protecting everyone in the water. A great side benefit is that the transformer automatically corrects polarity problems from the shorepower. Reversed polarity can be dangerous because AC appliances that should be off when their power switch is turned off will still have current flowing into them. Even worse, when polarity is reversed on some household appliances, such as refrigerators, the metal case may be energized with 120VAC. Anyone who comes into contact with that refrigerator and a ground could be electrocuted. Isolation transformers also prevent galvanic corrosion that can occur between boats in a marina that share a common ground through the AC shorepower. This connection can cause neighboring boats to damage or destroy each other's less noble underwater fittings, like aluminum outdrives. And finally, isolation transformers supply clean power to such sensitive AC electronics as computers and plasma TVs.

OFF GRID SOLAR POWER

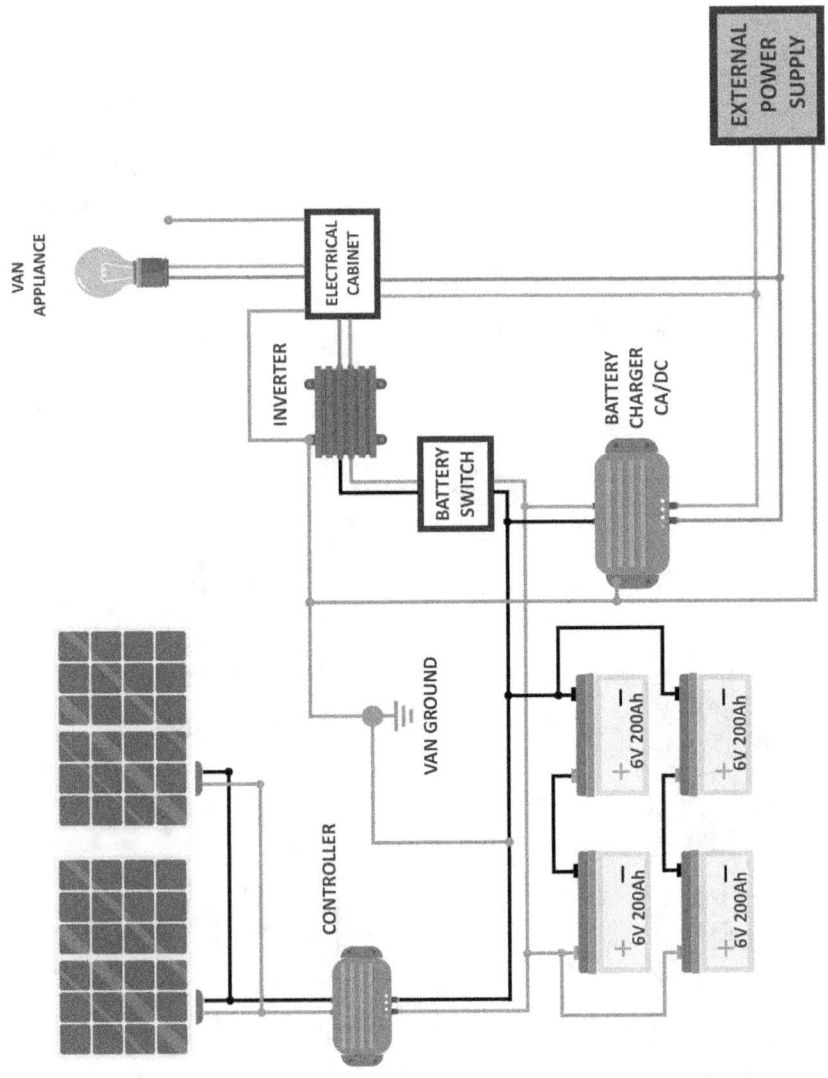

Schematic for RV with external power supply.

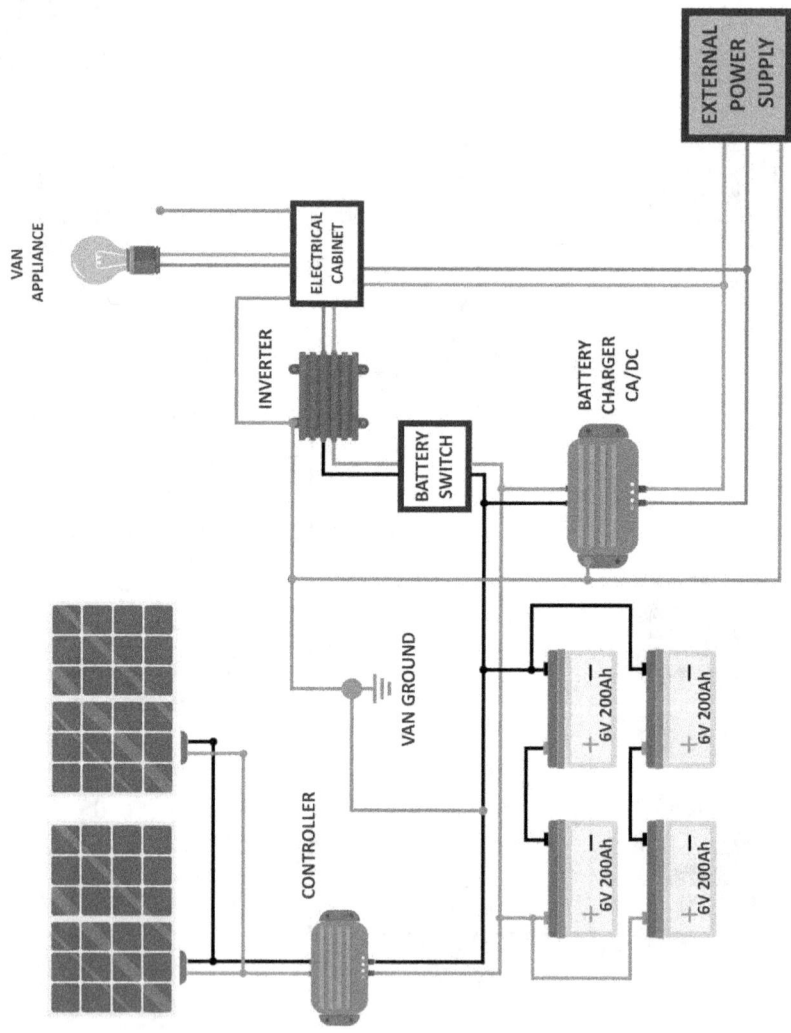

Chapter 4: Tips and Common Mistakes When Building an Off-Grid Solar System

Building your off-grid solar power system will be life-changing. Sometimes I believe people underestimate the number of adjustments that this lifestyle requires. If you have read until this point, you have demonstrated some level of curiosity and resilience - both being the necessary qualities to possess to fulfill this DIY project successfully. Your mindset will influence how much preparation you will undertake before building the system and how determined you are to understand terminology and concepts that may have been foreign to you before you learned of them in this book. In this last chapter, I want to prepare you with tips and suggestions that will help make this journey a meaningful one, resulting in a successfully installed off-grid solar power system. Below are a few tips that you should always remember when building solar power systems.

Tip 1: Do not go above 100 amps.

I always prioritize safety, not only in protecting yourself but also in building safe solar systems. There are so many dangers and problems that could arise when you use an extremely high electrical current. A simple mistake made in a highly charged system can be deadly or lead to a property fire. For safety reasons, most solar charge controllers will instill a maximum of 80 amps, and you will also find that deep cycle solar batteries avoid providing a high current. It should be reassuring that your system components are regulated however it is your responsibility to be cautious in how you use and wire the components.

Tip 2: Use the highest voltage.

It is common to find many people opting for a 12-volt system because they believe it will be strong enough to power their entire set-up.

However, in reality, when using 12 volts, you can experience a lot of voltage drops. Voltage drop refers to the decrease of electrical power along the path of a current, which flows in an electrical circuit. Solar batteries don't usually provide a high current and inverters have a specific voltage range they will operate within. Therefore, when the battery bank voltage drops too low, the inverter will simply shut down. The solution to this problem would be to increase the system voltage either from 12V-24V or 24V-48V.

Tip 3: Purchase the largest wires that you can afford.

Another way to reduce the risk of voltage drop is to purchase the largest and fattest wires that money can buy. This is because heavier wires carry less resistance than smaller and thinner wires. Furthermore, having heavier and larger wires allows you to carry greater loads or upgrade your system in the future. This will enable you to plan and avoid purchasing or replacing more wires a year or two down the line (unexpected costs are not fun).

Tip 4: Overbuild, don't underbuild your system.

There are so many parts, considerations, and processes to account for when constructing an off-grid solar power system. The accuracy in determining how much power you need in your homes is crucial in building a functional system. It is so easy to miss a calculation or overlook certain figures, which ultimately results in an underperforming system. Therefore, I would recommend you add at least 20 to 30 percent more solar panels and batteries than what you had already calculated, to give your system room to expand if needs be.

Tip 5: Wash your panels regularly.

I always recommend that people place their solar array in an area within their reach because they will need to have direct access to them regularly. Our solar panels are exposed to a lot of dirt, dust, bird poop, leaves, or pollen, and this - if not frequently removed - can decrease the power output of the solar panels. A rainshower will not do a great job of cleaning the panels for you; you need to personally wash them every now and then. To wash panels, take a non-abrasive brush and warm soap water and gently clean and dry panels throughout the year.

Avoid These Ten Costly Mistakes

There are more pros than cons in building your off-grid solar system. Indeed, there is a process which to follow and many factors to consider; however, the time and money invested in the initial stage will pay off. Now that I have shown you the steps involved in compiling your system, it is important to discuss some of the common mistakes that many DIY installers make when connecting their solar systems. Avoiding these mistakes will help individuals save a lot of time and money along the way.

Mistake 1: Incorrect solar system sizing.

Incorrect sizing of a solar power system is the most common mistake to make. Most of the time, it is due to mental shortcuts and basing the power requirements on factors that do not accurately represent how much power is used. For instance, you may look at your last utility bill and conclude that you use X amount of power when, in reality, this amount fluctuates throughout the seasons. There are many other factors that you must consider when deciding on power usage, such as changing climate temperatures, the positioning of panels, panel efficiency, and so much more.

Mistake 2: Leasing your system.

Solar power systems are only a sound investment if you own the system. Leasing a solar system presents many disadvantages. Firstly, the region offers tax incentives for installing solar systems in homes, by leasing your system you would not receive these incentives (they would go toward the real owner). Secondly, leasing may seem more affordable on a month-to-month basis however you will pay an excessive amount of interest. It would also tie you into contracts that directly oppose your lifestyle of being self-sufficient. Therefore, I would suggest that you purchase your own system and give yourself the peace of mind that you deserve.

Mistake 3: Not thinking ahead.

When many decide on building an off-grid solar power system, they make the mistake of calculating how they use power in their homes now and fail to think about how the power will be used in future homes. It would be an expensive endeavor to have to expand your system later in life because of unforeseen life circumstances. Therefore, when you plan for the system, consider if you are expanding your family, building a workshop, or purchasing electrical devices or machines that require more power in the future. Moreover, ensure that there is sufficient space around the area where to mount the solar arrays for future expansion. Consider purchasing batteries that are large enough to carry more power in the future (note that lithium batteries can be expanded in the future, but lead batteries are limited).

Mistake 4: Buying cheap solar panels trying to save money

I completely understand why a discounted price of a solar panel would be appealing for any customer (I love a good discount too). However, you need to know when to draw the line between a discount or a poor quality solar panel being dumped into the market. Due to the rise in demand for solar power, there have been some new entrants into the industry and many offering systems that are made in China at eyebrow-raising prices. The first point I want to stress is that if the price seems too good to be true, I guarantee you that it is probably true. I have seen some great, affordable solar products and some with pricing that was not feasible. Secondly, I always suggest individuals get at least three quotes from three different solar product suppliers regarding purchasing their solar system components. Compare the price as well as any extra services or warranties that the products come with (for example, receiving free customer support would be a great incentive).

Mistake 5: Overlooking the warranty terms.

Within the solar industry, it is believed that the length of a product's warranty is directly related to the manufacturer's confidence in their product. In other words, this means when the warranty is short, it is assumed that the product is of low quality and results in poor performance. Adopt this same mentality when purchasing solar products and aim to purchase products with the longest amount of protection. Understand what the warranty entails and what services it covers. Please do not be fooled by performance warranties which will provide the general lifespan of a product throughout the industry. Seek a manufacturer's warranty which will usually last for about 10 years.

Mistake 6: Using sun-tracking systems.

Many individuals believe that using sun-tracking systems will assist them in increasing the quantity of energy that their solar panel array produces. However, in reality purchasing a sun tracking system may be just an extra expense on the ever-growing list. This is because you are better off purchasing more panels for your system and being proactive about your need for more energy production instead of the passive-route of purchasing a tracker (which may or may not help "track" more sunlight). Moreover, sun-tracking systems are prone to system failures, so don't expect it to last as long as your installed solar panels.

Mistake 7: Deciding not to purchase a battery monitor.

Battery monitors are extremely useful devices that provide much needed information about the status and health of the battery bank. This device helps maintain the bank by detecting issues presented in the off-grid solar power system or current levels of power available. This information is particularly useful to those individuals who are still learning more about solar power and need as much support in understanding their solar power system. Additionally, the battery monitor will measure and keep track of the total amp hours accumulated in the system, which subsequently allows you to monitor a household's energy usage.

Mistake 8: Not switching to more efficient electrical appliances.

Once you have a fully functioning off-grid solar power system, the last thing you want to occur is the misuse or waste of energy. Energy is wasted when you use appliances that consume a lot of needless energy. Switching lifestyle habits and deciding on using more energy-efficient appliances will help you conserve as much energy as you can while still enjoying the benefits of using electronic devices and appliances. A simple lifestyle adjustment, such as purchasing LED light bulbs instead of incandescent light bulbs, can save a lot of energy.

Mistake 9: Not investing in a backup generator.

Generators may seem like an unnecessary expense when setting up an off-grid solar power system, however they form an indispensable part of the system, especially when you have decided to go completely off the grid. One of the major reasons why one decides to go off-grid is having the freedom to power their homes in the manner that they desire. However, you cannot control unpredictable weather conditions but you can plan for that time. Your generator will be a lifesaver during worst-case scenarios, which can happen at any time. You cannot generalize when planning your off-grid solar power system and believe that the climate will always remain the same and neutral.

Mistake 10: Not staying up to date with new technology.

The solar power industry is still fairly new and as a result, there are always new updates and products released in the market. This is good news for you because it means you are always presented with opportunities to make the solar power system more efficient. It also means that it is becoming more affordable to own a solar system because of the demand of solar products. Keeping up-to-date with new technologies will help you purchase more efficient components, creating an overall efficient solar power system. It will also expose you to more environmentally-friendly alternatives of building a solar power system which will reduce your overall carbon footprint.

Maintaining Your Solar System in Every Season

One of the things I love about solar panels is that they are extremely reliable. Once you have grounded your system in place, the parts will never move for its entire lifespan of 25 to 30 years. Some suppliers do offer warranties that come with maintenance services as part of the purchase; however, not every supplier will do this. Most of the time, you will have to carry out preventative maintenance for your off-grid solar power system in order for the system to remain as efficient as possible. These maintenance measures can be carried out in every season of the year, or depending on the climate in your particular geographical location.

In general, I would say that solar panels are quite low maintenance, and when there are present failures in the solar power system, the failures are usually related to the electricity production or the corrosion of wires that connect the system to the inverter. This means that the solar panels specifically will not give you too much of a hassle (this is great news indeed). Nevertheless, don't forget that the panels can only assist in producing electricity when there is sunlight. This means that any obstructions, such as shade or gloomy weather conditions, will reduce the amount of energy production that the panels can generate (and in extreme cases halt energy generation completely). Below are some maintenance tips you can follow to care for your solar power system in every season.

Summer Maintenance

Most homeowners love the summer because it is an extremely productive season for solar panels. Nonetheless, homeowners need to be mindful of the dust, pollen, or animal droppings that will land on the solar array every so often. You also cannot rely on summer rain showers to do the cleaning for you; they will not do a thorough cleaning the way you can. A simple rinse with warm soapy water and a non-abrasive brush should be enough to keep panels clean and efficient. Those individuals who live in areas where there is a lot of dust in the air during summer need to check on their panels regularly and simply rinse them off with a garden hose whenever necessary.

Fall Maintenance

Solar panels are durable enough to continue generating energy during the fall months. This, of course, is only possible when the panels are not covered in leaves, dust or dirt. There is no way to keep the leaves away from interrupting the power system, even when there aren't any huge trees near your home. Your greatest challenge during fall will be the autumn winds that carry debris and other vegetation in the air and offload them on panels and racking systems. Therefore during the fall months, your task will be to remove the leaves that may obstruct the panel's ability to receive full sunlight. To do this, you need equipment such as a ladder, rope, and a firm harness to help prop you on to the roof (if panels are mounted on the ground, this process will be much easier). After that, use a soft brush to rake the leaves off the panels or use a leaf blower.

Winter Maintenance

Many homeowners fear that their solar panels will not produce much, if any, electricity. While it may be challenging to receive full days of sunlight during winter, your solar panels will still absorb as much as they can take. There will be days (even when it is a half-day) where the sun will come out, and on these days ensure there are no obstructions such as snow on panels which can sabotage power production. It may be a struggle to remove snow on the solar panels, especially in areas where there is a high chance of snowfall every week during the winter.

In cases like these, you will need to rely on the sun to melt the snow away within a day or two of it landing on the panels. However, for the snow to melt away, make sure to mount panels with at least a 15-degree tilt. Be cautious of ice build-up because it can compromise your roof's integrity, which will ultimately impact the structure of the solar power system. I do not recommend that individuals apply salt as a method of melting the snow because it could potentially damage the systems, causing the erosion of the racks and panels. By all means, try to remove as much ice as possible manually.

Spring Maintenance

I find that spring is the best time to do preventative maintenance work to expand the lifespan of your solar panels. In this time, you can focus on conducting an inspection of the installation, wiring and the health of the inverter and battery bank to address any prevalent issues. The Solar America Board for Codes and Standards has compiled a list of things to observe when conducting an annual solar power system inspection. These include, but are not limited to: water leakage on the roof, roof drainage inefficiencies, growth of plants near the panel installation area, corrosion on the racking system or on electrical enclosures, signs of pest infestation, cracked glass on the panels, loose or missing bolts, and excessive wear on the inverter.

Conclusion

Solar power systems have revolutionized how people live and work. They have shown us ways of living more efficient lifestyles and working in greener ways; they have also allowed us to see how simple it is to live with less and yet have a full and meaningful life. Our consumerist society taught us the reverse; their message was centered on acquiring more in order to have a meaningful life. The culture of "more" has resulted in more spending, more bills, and more debt. Thankfully, many people have started to open their eyes and see how more is not always beneficial. This realization has led to a new wave of people opting to reduce their financial obligations and free themselves from crumby creditors or utility companies whose profit-motive is to continuously raise the cost of accessing electricity. Perhaps it was to our own benefit that these companies raised the cost of accessing electricity because if we did not incur such a high expense, we would not have discovered or considered solar power.

Solar power was not always popular or widely used in our society. In the past we did not have access to another environmentally conscious method of generating electricity. Our only hope for receiving electricity in our homes was in burning fossil fuels and praying that the reserves of these fossil fuels would not deplete in our lifetime. Not only are fossil fuels creating more air pollution in our environment through the release of carbon dioxide, they are also seen to be unsustainable. It is due to their unsustainability that the price to access and produce energy from fossil fuels has dramatically increased overtime. Paying exorbitant fees for a volatile non-renewable energy resource does not seem to make any sense, especially with the launch of solar products that offer consumers a sustainable renewable energy resource which can power their homes for a once-off investment.

Admittedly, when solar power systems and other products first came out in the market, they were seen to be expensive toys for the super-rich, which they could experiment with in their large homes. However, due to the demand for solar power as a formidable alternative to fossil fuels used for generating electricity, the market has become more accessible to more people. Already as it stands, the price tags of solar products have dropped considerably. This trend will continue to gain momentum as newer and more efficient cost-saving technologies are introduced. The democratization of the solar power industry has made it possible for those who have the courage to live off the grid, to own one of these durable and reliable systems, and ultimately freeing them from depending on greedy utility companies.

If it were up to me, I would give a special discount on solar systems for all remote homeowners. This is because this kind of technology offers them an unprecedented amount of freedom to live a self-sufficient life with all the perks of having electricity and heating within their home. Going off-grid is a commendable lifestyle decision, and I believe that the experience is heightened when the home can generate its own electricity. Unlike purchasing power from a utility company that would keep a household under a life-long contract, an off-grid solar power system provides the homeowner with the independence to use X amount of electricity to power X amount of items. Moreover, this gives the homeowners freedom to determine how they desire to use energy.

The process of setting up an off-grid solar power system is a straightforward one to follow. In this book, I have shown you the list of components that make up an off-grid solar power system and how each element is useful in turning solar power into usable electricity. I believe one of the relevant messages of this book is that each household will have its custom system dependent on their lifestyle. In other words, the solar power system will depend on how large your household load is. The greater the number of electrical appliances, devices, and machinery you use on a daily basis, the greater the off-grid solar power system will need to be. Another important factor related to the system's size is that you should always think ahead and plan your system after having factored how living arrangements and lifestyles would be adjusted in the future.

Planning an off-grid solar system based on how you are living today would result in having to expand in the future. These expansions would only cost more time and money to accomplish. My hope for you is that you can build a sustainable system that can survive for decades with low maintenance as your only regular task. The maintenance work can be done throughout the year, with the main priority being to remove any obstruction or dirt, reducing the energy production of the panels. As I have mentioned in the book, once your solar array has been correctly fitted, the solar panels will do the necessary work of converting energy into electricity without requiring assistance or supervision from you. As long as you build a system with a good enough charge and with wiring that is safe and functional, you can begin to enjoy the benefits of solar electricity almost immediately.

I also discussed at length about the cost factor involved in regards to owning and installing an off-grid solar power system. Indeed, these kinds of systems can be expensive; however, when we consider the once-off cost and compare it to the cost of staying on the grid, we will soon realize that the price tag is more affordable than what we may have thought. Solar panels are an extremely low-maintenance device that does not require a lot of handling to function optimally for decades at a time. When we experience problems in the system, it usually has to do with wiring that may have eroded or batteries that have worn out. These repairs are preventable when you invest in large wires and durable batteries when first setting up the system.

Your Gateway to Freedom

An off-grid solar power system is worth the capital investment it requires because of the immediate freedom it offers us. In all truthfulness, we deserve to live a life free from being tied under any contractual obligation. Our money can now serve us and be invested in technologies that will serve our needs instead of keeping us in bondage to others. Off-grid solar power systems celebrate a lifestyle of being connected to the environment in ways that you could have never experienced living on the grid. The power of the sun becomes more profound when you realize how generous this ball of gas is in providing solar energy for us to use freely. Furthermore, I always find that those who live off the grid are more conscious of wastage and so they seek by all means to store food, water, and electricity that they do not use for later consumption.

This new off-grid lifestyle supports autonomy even in learning skills that are necessary for survival. I hope that this book has taught you a new skill that will help you further establish your autonomy by building an off-grid solar power system. I have faith in your ability to consider all of the necessary calculations to design an efficient and functioning system. It doesn't really matter whether you live in a tiny house, cabin, RV, or houseboat, the off-grid solar power system design is generally similar in how it is connected in all types of small homes. The only factor which will be different in all small homes is the household load that the solar power system is required to carry—smaller household loads will result in smaller solar power systems and vice versa.

Another benefit of learning how to build and install your own solar power system is that you can save thousands of dollars required to hire a professional. Now that you understand the calculations and the various assumptions and considerations that you need to make when designing your off-grid solar power system, you certainly have the skills to construct and connect the system on your own. All it takes to install a functional solar power system is an individual who can understand how the system works and how each component carries its own charge, which will in essence affect the system's overall charge. The guidelines shared with you in this book are the perfect blueprint to help you design your own system with the least amount of errors or confusion experienced in the process.

It is your responsibility to set yourself free and live a completely self-sufficient lifestyle off the grid. I believe that in this book, I have given you all of the fundamental tools and guidelines in making your DIY solar project a successful one. Do not be afraid to ask for guidance along the process of building your off-grid solar power system from suppliers or manufacturers. Furthermore, I would recommend that you purchase solar products that come with warranties so that you are able to access free maintenance or repair services for a particular period of time. Finally, it would even be good to have an extra pair of hands helping you lift the racking system, drill bolts in, dig holes and install a solid structure for your solar array. Perhaps you can commission the help from a DIY-loving friend or relative?

I believe that this book should be the beginning of your research into building your solar power system—not the end. There is so much more information available to you online and through forums that will help you master a skill that only a few individuals know how to do successfully. I implore you to take all of the valuable knowledge that I have shared with you in this short amount of time and continue finding ways to make your system even more efficient. Once again, stay updated on the latest technology entering the market and assess their strengths and weaknesses; this will help you to make brilliant purchasing decisions when it is time for you to buy your own off-grid solar power system.

Most of the information shared in this book will be new to you however, do not let this intimidate you. Embrace the unknown. Stretch your mind a little and challenge yourself to assume a greater amount of responsibility for how well you live. As much as there is a lot to complain about in our society, there are also so many alternative ways of living which can free us from systems that seek to keep us in bondage. Your tiny home was the first step in liberating yourself from the familiar fast-pace urban lifestyle that brought much discontent. Your final step is to build your own system which will help you generate your own electricity and kiss the life of dependency goodbye for good.

References

8 Costly Solar Mistakes to Avoid When You Design Your System. (2018, August 13). Wholesale Solar. https://www.wholesalesolar.com/blog/solar-mistakes

2020 Off-Grid Solar System Cost | How Much Does It Cost To Go Off-Grid With Solar? (n.d.). Fixr.Com. https://www.fixr.com/costs/off-grid-solar-system

A Guide to Off Grid Solar | Green MagazineGreen Magazine. (n.d.). Green Magazine. https://greenmagazine.com.au/information/solar-calculator/a-guide-to-off-grid-solar/

BBC. (n.d.-a). *Fossil fuels - Fossil fuels and nuclear power - GCSE Physics (Single Science) Revision - Other.* BBC Bitesize. https://www.bbc.co.uk/bitesize/guides/z2d2bk7/revision/1#:~:text=Disadvantages%20of%20using%20fossil%20fuels&text=Fossil%20fuels%20release%20carbon%20dioxide

Beaudet, A. (2015, October 2). *First Steps to an Off The Grid Solar Power System | altE Blog.* Solar Power News & DIY Solar Tips. https://www.altestore.com/blog/2015/10/first-steps-for-planning-an-off-the-grid-solar-power-system/#.XxrXPZ4zbIW

Casey, B. (2017, May 8). *Essential Tools for Solar Installations.* Solaris. https://www.solaris-shop.com/blog/essential-tools-for-solar-installations/

Conde, J. (2012, December 10). *PV Installation Tools and Techniques - Solar Novus Today.* Www.Solarnovus.Com. https://www.solarnovus.com/pv-installation-tools-and-techniques_N6011.html

Coulee. (2020, March 29). *11 Top Tips for Going Off-Grid Solar Systems.*

Coulee Limited. https://couleenergy.com/off-grid-solar-systems/

Damon on Road. (2019a). *Oulaw RV Parked on Grass Near Trees.* In *Unsplash.* https://unsplash.com/photos/6q3falrsdO0

Dilthey, M. (2017, March 14). *How to Maintain Your Solar Panels Year Round.* Solar Power Authority. https://www.solarpowerauthority.com/maintain-solar-system-year-round/

D.Light. (2019b, March 11). *Benefits of Off-Grid Solar System [Infographic] | d.light.* D.Light. https://www.dlight.com/blog/benefits-of-off-grid-solar-system-infographic/

Fovre, T. (2019). Blue and Gray Solar Panel. In *Unsplash.* https://unsplash.com/photos/Ac_6K-ELfJg

Ivan, & Manuela. (2019, September 9). *How to Wire a Tiny House for Solar Power.* Tiny House Bloom. https://tinyhousebloom.com/how-to-wire-for-solar-power/

Mozaw. (2018a, June 13). *Design and Build Off Grid Solar Power System: DIY Complete Guide Step-by-step.* Green Living Blog. https://mozaw.com/diy-off-grid-solar-system/

Narasimha, V. (n.d.). Solar Water Heater Solar. In *Pixabay.* https://pixabay.com/photos/solar-water-heater-solar-water-331316/

Northern Arizona Wind & Sun. (n.d.-b). *RV Solar Electric Systems Information.* Northern Arizona Wind & Sun. https://www.solar-electric.com/learning-center/rv-solar-electric-systems-information/

ØRBÆK, A., DAHL, J., & AMADEI, C. A. (n.d.). *Design of an off-grid Photovoltaic system With supplementing energy from Wind and Diesel.* https://scholar.harvard.edu/files/amadei/files/off-grid.pdf

Peacock, F. (2011, December 14). *Dont Make these 10 Mistakes When*

Buying a Solar Power System. Solar Quotes Blog. https://www.solarquotes.com.au/blog/dont-make-these-10-mistakes-when-buying-a-solar-power-system/

PhotoMIX. (n.d.-c). Solar Panels Heating. In *Pixabay.* https://pixabay.com/photos/solar-panels-heating-1477987/

Pixabay. (2015). Brown Grey Wooden House Near Lake at Daytime. In *Pexels.* https://www.pexels.com/photo/brown-grey-wooden-house-near-lake-at-daytime-158316/

SEIA. (2019c, May 9). *United States Surpasses 2 Million Solar Installations | SEIA.* SEIA. https://www.seia.org/news/united-states-surpasses-2-million-solar-installations

Seious Young Man Working Near Solar Panels. (2020). In *Pexels.* https://www.pexels.com/photo/serious-young-man-working-near-solar-panels-4254159/

Solar 101: How Solar Energy Works (Step by Step). (2019, May 2). CertainTeed. https://www.certainteed.com/solar/solar-101-abcs-solar-power/

Solar For Tiny Houses: What You Need To Know | EnergySage. (2019, February 8). Solar News. https://news.energysage.com/solar-for-tiny-houses/

SOLAR MISTAKES - Smart Quotes Australia. (n.d.). Smartquotesaustralia.Com. https://smartquotesaustralia.com/solar-mistakes/

The NEED Project. (2018b). *Solar What Is Solar Energy? Fusion Solar Space Heating.* https://www.need.org/files/curriculum/infobook/solari.pdf

The Pros and Cons of going Off the Grid. (n.d.). Carbontrack. https://www.carbontrack.com.au/blog/off-grid-pros-cons/

Vladvictoria. (n.d.-d). Boat Ocean Sailing Vessel. In *Pixabay.* https://pixabay.com/photos/boat-ocean-sailing-vessel-leisure-

4001849/

When Is A Ground Mount Preferable To A Rooftop Solar Power System? - YellowLite. (2016, July 27). Www.Yellowlite.Com. https://www.yellowlite.com/blog/post/when-is-a-ground-mount-preferable-to-a-rooftop-solar-power-system1/

www.ingramcontent.com/pod-product-compliance
Lightning Source LLC
Chambersburg PA
CBHW050256220526
45465CB00002B/710